AF320539

Marat Khaitbaev

# Modern energy supply in the automotive industry

**Marat Khaitbaev**

# Modern energy supply in the automotive industry

**ScienciaScripts**

**Imprint**

Any brand names and product names mentioned in this book are subject to trademark, brand or patent protection and are trademarks or registered trademarks of their respective holders. The use of brand names, product names, common names, trade names, product descriptions etc. even without a particular marking in this work is in no way to be construed to mean that such names may be regarded as unrestricted in respect of trademark and brand protection legislation and could thus be used by anyone.

Cover image: www.ingimage.com

This book is a translation from the original published under ISBN 978-620-8-11611-8.

Publisher:
Sciencia Scripts
is a trademark of
Dodo Books Indian Ocean Ltd. and OmniScriptum S.R.L publishing group

120 High Road, East Finchley, London, N2 9ED, United Kingdom
Str. Armeneasca 28/1, office 1, Chisinau MD-2012, Republic of Moldova, Europe
Printed at: see last page
**ISBN: 978-620-8-12193-8**

Copyright © Marat Khaitbaev
Copyright © 2024 Dodo Books Indian Ocean Ltd. and OmniScriptum S.R.L publishing group

"For my Julia."

- Marat Khaitbaev

# Table of Contents

# Introduction

The concept of smart home and smart transport was initially based on the use of solar energy to supply all the energy needs of its infrastructure elements.

These solutions have been demonstrated at worldwide solar energy exhibitions and have largely involved the use of solar-powered devices to recharge the batteries of electric vehicles.

Already in the first demonstrations, it was identified that in the smart home and in smart transport, batteries and accumulators are needed to store and store electricity while minimising its consumption in order to ensure a stable supply of electricity.

The first experiments on the use of car batteries and accumulators for this purpose clearly showed the acute dependence of storage devices on heating and forced the search for a clear solution to this problem through the use of new construction materials for the manufacture of housing parts of energy storage devices and accumulators intended for use in infrastructure and smart home subsystems with the possibility of subsequent use in electric vehicles.

## Infrastructure and ecosystem for smart home and smart transport

The book deals with innovative infrastructure components of smart home and smart transport in terms of efficient solution of energy supply problems and application of the most advanced integrative solutions based on the use of the latest composite materials for solving these and similar problems.

As a result of a deep search it was determined that to solve the problem of heating of electrical installations it is necessary to use a composite material, which is simultaneously a conductor of electric current and an effective heat conductor, having a developed three-dimensional conductive structure, with evenly distributed nodes in it (most expediently - microspheres), with points of maximum thermal conductivity evenly distributed over the volume of the material, at the same time not being conductors of electric current (i.e. made of material with maximal thermal conductivity).

The material should thus have the form of a three-dimensional micro-lattice, in the nodes of which diamond spheres, which are the best known heat conductors, are arranged, separated in the three-dimensional space of the structure from each other by copper spherical shells, which are excellent conductors and heat conductors.

Thus for electric current (most importantly for current in pulse mode) the composite structure is a kind of pseudo-spongy or pseudo-porous volume, since throughout the said volume of the conductive material, dielectric spherical spaces, commensurate in size with the dimensions of the conductive space, are uniformly distributed.

This fact favours a rather rapid and uniform current dissipation on the one hand and a rapid, efficient, uniform heat dissipation on the other hand, phenomena occurring in the same volume of material.

**Figure 1: Models of a diamond sphere (red colour) and a copper shell (gold colour).**

The most ductile known materials, e.g. copper or silver, which have the highest electrical conductivity of any known material, are used as sheathing materials. When subjected to high pressure in a closed volume, these metals can be brought to a cold-fluid state.

When high pressure is applied in a three-dimensional closed volume, the nature and form of interaction between the capsules in the structure are modified, which makes it possible to form products with the necessary technical and technological conditions that cannot be obtained with conventional technologies.

The new material can obtain its unusual properties due to appropriate technological techniques, which, with their originality, become basic for a complex technological process - the object of the basic invention and a series of applicative inventions aimed at developing and improving the properties of said composite materials and their derivatives.

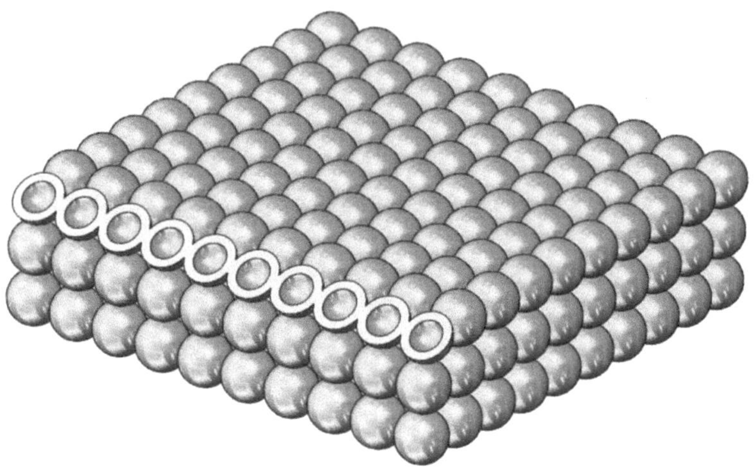

**Figure 2**: **Three-dimensional model of the composite material.**

Taking into account all the preliminary characteristics of the proposed construction material for the manufacture of body parts, it is possible to give it a more complete characterisation, which at this stage of analytical processing can be a formulation and at the same time a technical characterisation.

# Description, characteristics, properties and advantages of the composite material

A composite modular material with high thermal conductivity, high electrical conductivity, capable of absorbing and dissipating significant amounts of energy in very short periods of time, absorbing and transmitting significant amounts of energy over a distance, and having maximum mechanical strength, with maximum reliability in maintaining accurate geometric shapes when exposed to high concentrations of temperature, energy and other types of harmful extreme effects.

In the subsequent stages of analysis, a more coherent material characterisation is drawn out, in which parameters are introduced to identify the steps and steps along the production process of this innovative material.

Formulation of a new composite material as a product:

- a composite material having a developed three-dimensional (volumetric) structure consisting of a set of identical multilevel spherical shells covering spherical cores. The cores with shells (capsules) are bound together through a number of consecutive technological operations and have an equivalent form of contact between each other for all capsules of the structure;

- The composite material has properties of super thermal conductivity and super electrical conductivity;

- composite material has high mechanical strength, is not prone to internal mechanical and thermal stresses and, as a consequence of these phenomena, to the occurrence of internal deformations;

- the composite material is capable of being subjected to high pressures and is capable of entering a cold-fluid regime under the influence of these pressures for at least part of the components, which makes it possible to calibrate the three-dimensional geometric shape of the structure and to ensure very precise geometric dimensions of the structure with a high degree of repeatability.

Variants of the name and definition of the composite material production technology:

A method for fabricating a pseudo spongy or pseudo porous composite material comprising a plurality of nanoscale capsules bonded together to form a three-dimensional structure subjected in the final stage of fabrication to volumetric plastic calibration deformation in a cold-fluid mode for a plastic shell material of the nanoscale capsules.

The technique of producing nanoscale powder from diamond and then coating it with copper or other plastic metals is a technique relatively well-known in terms of technology principles, but in the later stages of the project, requiring relative modification.

As a result of an in-depth patent search and comparative analysis of its results, the main distinguishing features of the proposed material are determined:

- conductive and dissipative of electric current;

- heat-conductive and heat-dissipative;

- conductive and heat-dissipative;

- heat conducting and dissipating electric current,

- generally regarded as complex integrative composite compounds:

- on the basis of microscopic spherical constituent components connected with each other in the process of realisation of cold-fluid principles,

- in the form of spherical multilayer capsules having a dimensional factor of all structural elements

- in the nanoscale metric range.

A three-dimensional, composite, pseudo-sponge structure comprising a plurality of layers having equivalent (identical) geometric shape and each consisting of at least two layered components in contact with each other and forming a complete three-dimensional geometric shape, wherein the said components are uniformly and equivalently distributed over the volume, have and form equal conditions of electrical and thermal interaction with each other, wherein the same-type layers of all components are separated from each other by a single layer.

A three-dimensional structure, characterised in that each layer at each component is a closed three-dimensional geometric figure.

A three-dimensional structure, characterised in that each successive layer at each component covers the entire surface of the previous layer at each component.

The objectives set out in this book are:

- increasing the power of electronic devices in which the proposed materials are supposed to be used;

- reducing the size of electronic devices in which the proposed materials are to be used;

- increasing the level of reliability of electronic devices in which the proposed materials are supposed to be used;

- lengthening the life of electronic devices in which the proposed materials are to be used;

- Improving the overall efficiency of electronic devices in which the proposed materials are to be used.

The proposed composite material is capable of fundamentally changing the operating conditions and performance characteristics of high-energy electronic devices and makes it possible to create a new generation of electronic devices that are much less dependent on thermal characteristics. This is especially important for high-power pulsed devices having power at the pulse peak greater than the rated power of the device.

As an example, a single-structure semiconductor laser with a nominal output optical power of 300 milliwatts and a wavelength of 780 nanometres, which, when connected to a control electronic module operating in the radio-frequency range (100 megahertz) at a pulse peak of 10 nanoseconds duration, repeated every 10 nanoseconds, showed an output optical power of 3.1 watts for 72 hours. The hetero-structure of the above semiconductor laser diode was mounted on a substrate made of the proposed composite material in the form of a pseudo-sponge structure.

Additional opportunities that the use of the proposed material provides:

- manufacture of instrument housings from the same material with a homogeneous monotonous structure;

- execution of cases and supporting parts of electronic devices in the form of conductive sponge system, capable in case of sudden peak current pulsations or sudden peak temperature pulsations in the shortest time to dissipate or accumulate an excessive part of the suddenly arisen energy load;

- the possibility to combine current-carrying and heat-carrying functions in one and the same structural element.

What is supposedly invented as a result?

- structure of a multilayer (multilevel) capsule;

- geometric shape of the multilayer (multilevel) capsule is a sphere;

- order of alternation of layers (levels) in a spherical capsule;

- the order and geometry of spherical capsules in the three-dimensional structure of the product;

- the technological principle of manufacturing the product;

- introduction in the manufacturing process - the operation of calibrating the geometric shape of the product, after the first pressing stage;

- calibration operation in a three-dimensional coordinate system;

- performing the calibration operation when the material state of the outer layer (shell) of the capsule is close to or equivalent to the cold-fluid state;

- removal of all cavities not filled with conductive material from the three-dimensional space of the product during calibration;

- formation of a pseudo-spongy structure in the three-dimensional space of the product, with the role of separating points in the said structure played by less plastic materials from those used in the capsule composite;

- utilising the sponge structure of the product to dissipate heat and current throughout the entire volume;

- use of the pseudo-sponge structure of the product to absorb (absorb) excess energy arising during peak moments of the pulse mode of operation of the product;

- use of the cold-fluid state to relieve internal stresses in the material and dimensional calibration in three coordinates simultaneously;

- a combination of materials in a hierarchy of shells of a spherical shaped capsule such that each successive layer is made of a less hard and more ductile material;

- combination of materials in the hierarchy of the core and shells of a spherical capsule in such a way that the core is always made of the hardest material of all the materials used to create the capsule;

- application as the main principle of calibration - preservation without deformation of the solid core of the sphere and maximum level of plastic deformation of plastic materials of the peripheral layers of the capsule sphere;

- application for high specific pressure calibration in closed three-dimensional space;

- application of the principle of uniform pressure distribution along all coordinates (axes) of a closed three-dimensional space;

- selection of thicknesses of plastically deformable layers so that the minimum layer thickness is greater than or equal to the capsule core diameter.

The main feature of the composite material formed from spherical multilayer capsules in two stages, of which the first stage provides for the spheres to be brought into contact on the outer spherical surface and such contact is point contact, and the second stage provides for the intermediate workpiece to be placed in a three-dimensional closed volume geometrically equivalent to the calculated shape of the product and extreme impact on the workpiece by means of high pressure propagating along all axes and coordinates of the said volume, and the level of specific gravity of the spheres in the volume of the plastic conductive material. In this case, the workpiece takes the form of a pseudo-sponge structure in which nuclei of semiconductor, ceramic, diamond or conductor with certain properties are uniformly distributed throughout the volume of the plastic conductive material.

Advantages of composite when used in electricity storage and battery storage in smart home and smart vehicle:

- super thermally conductive, super electrically conductive, pseudo sponge composite three-dimensional structure providing maximum heat dissipation, maximum current absorption, low electrical resistance, low thermal resistance, low level of current losses during its passage through the structure, maximum speed of pulse signals passage, with minimum energy losses, maximum level of absorption of energy pulses occurring at high frequency and having a short duration comparable to the frequency of the pulses.

Indirect benefits include the following:

- materials and nanoscale spheres for use as capsule cores are mass-produced based on several identical manufacturing processes;

- technological processes for the application or construction of layers (shells) following the core are known and tested;

- Volumetric calibration processes are used in cold extrusion techniques for the production of moulds, dies, etc.

The proposed technical solution may become the subject of invention. Preliminary search works have been carried out.

In order to assess the real innovation potential, the author has performed preliminary design and construction work to tie into a Porsche classic model passenger car without going into detail, but only in principle.

# Scope of application of composite materials

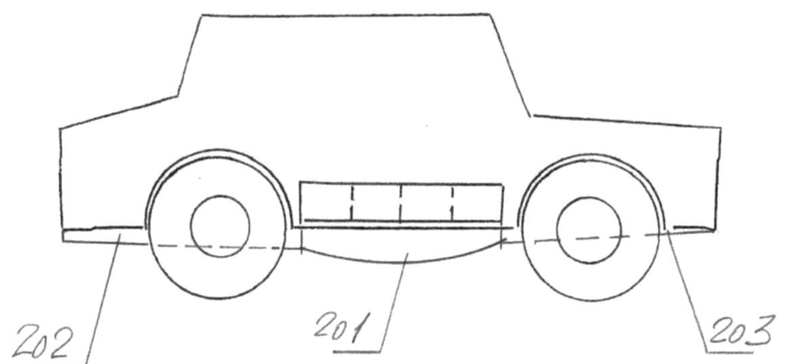



Figure 3

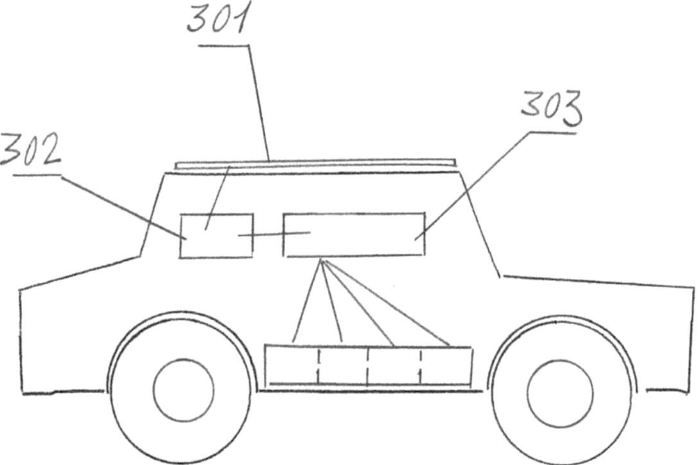



Figure 4

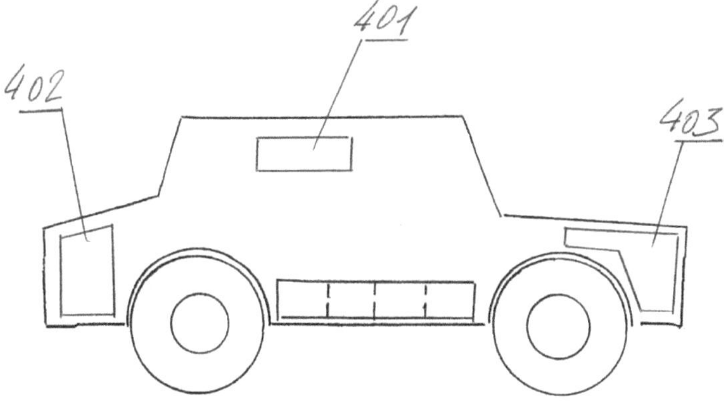


**Figure 5**

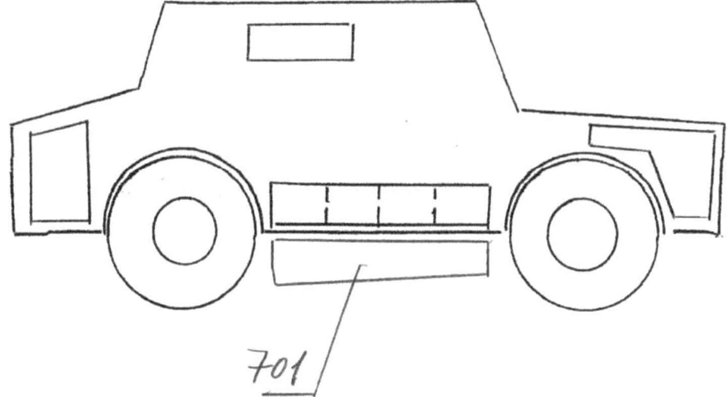


**Figure 6**

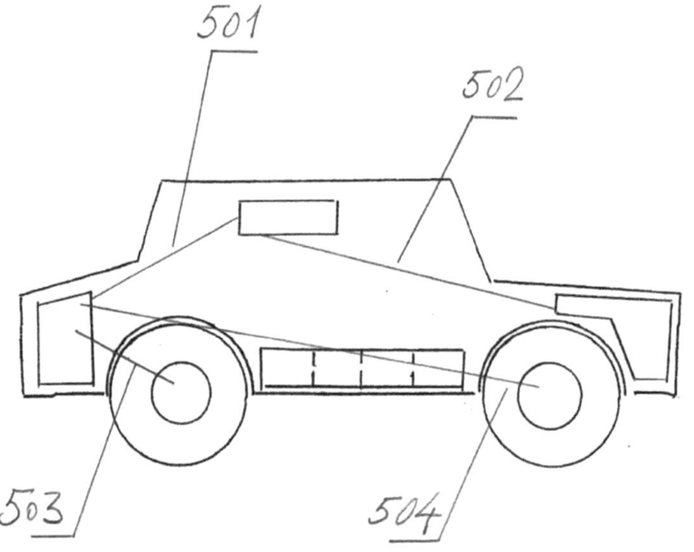


**Figure 7**

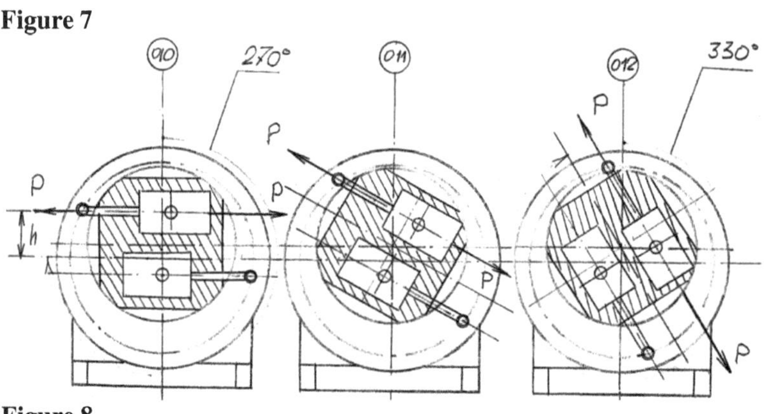


**Figure 8**

16

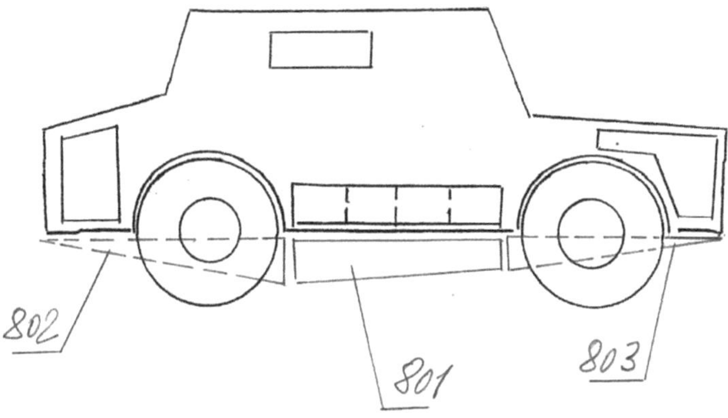


**Figure 9**

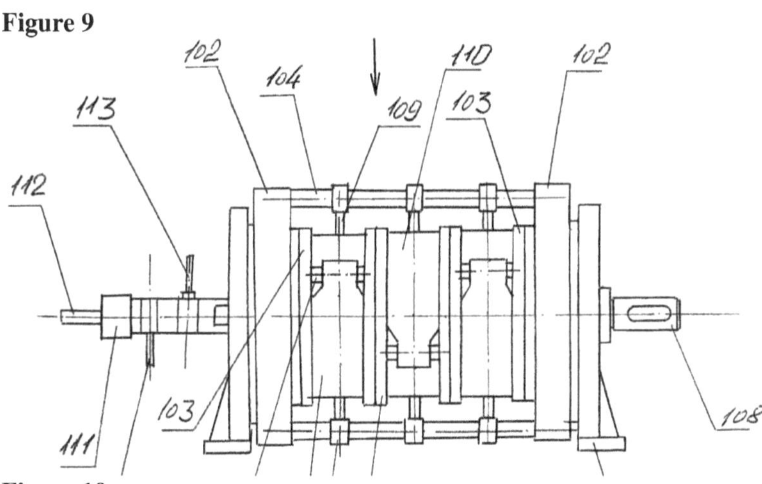


**Figure 10**

17

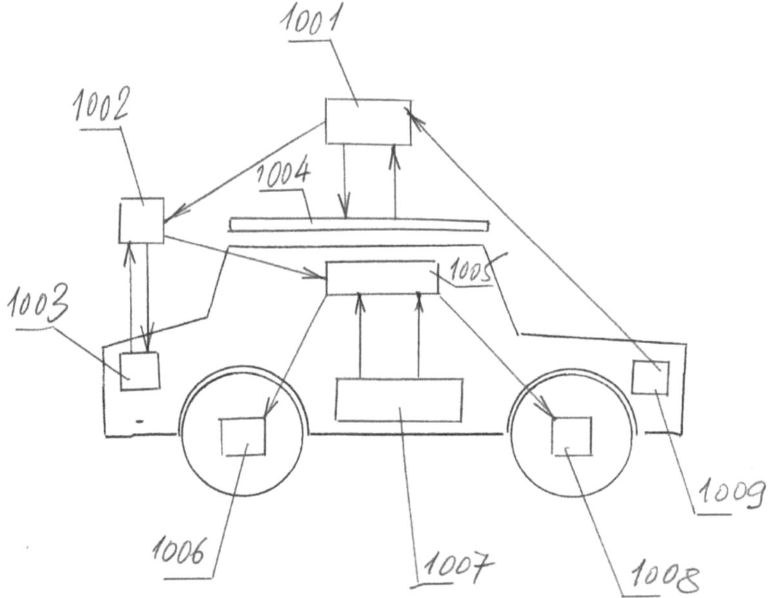


**Figure 11**

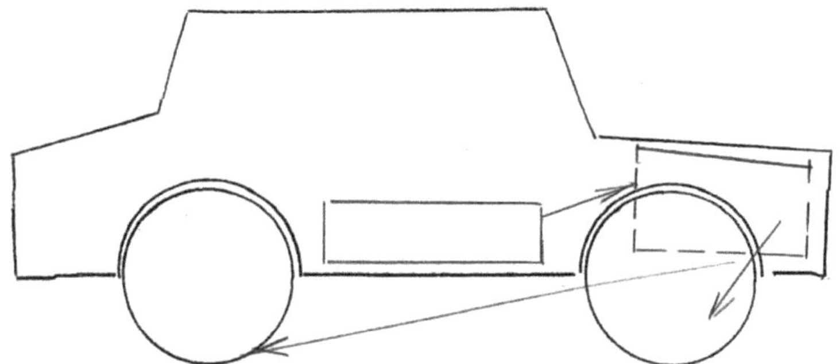

**Figure 12**

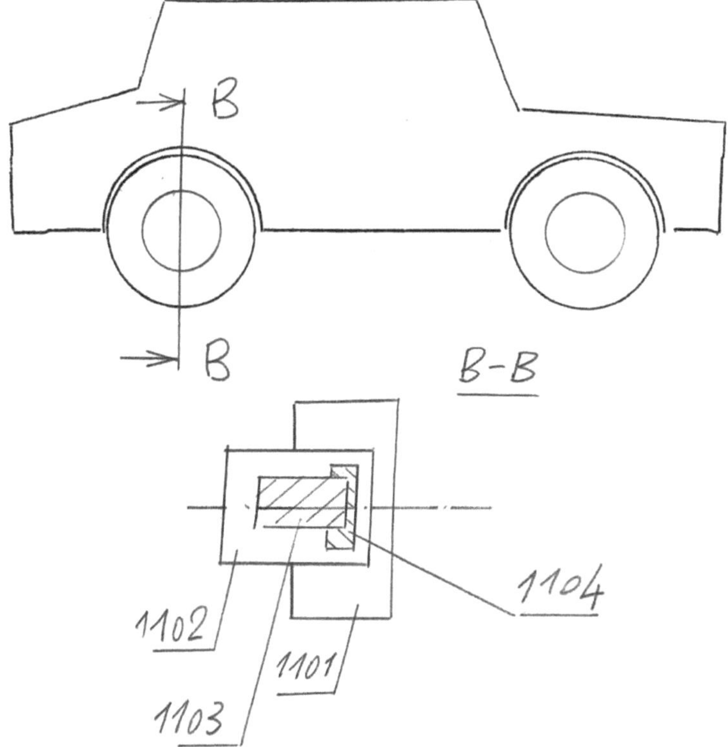


**Figure 13**

Figures 3-13 show examples of step-by-step development of models for modernising passenger cars to a hybrid engine structure and to an electric vehicle structure with the possible use of electric batteries with housings made from the composite materials being developed.

The design elements are arranged in a way that allows for brainstorming processes.

The drawings include elements of kinematic diagrams for possible development options and rotary engine versions, and in an innovative aspect.

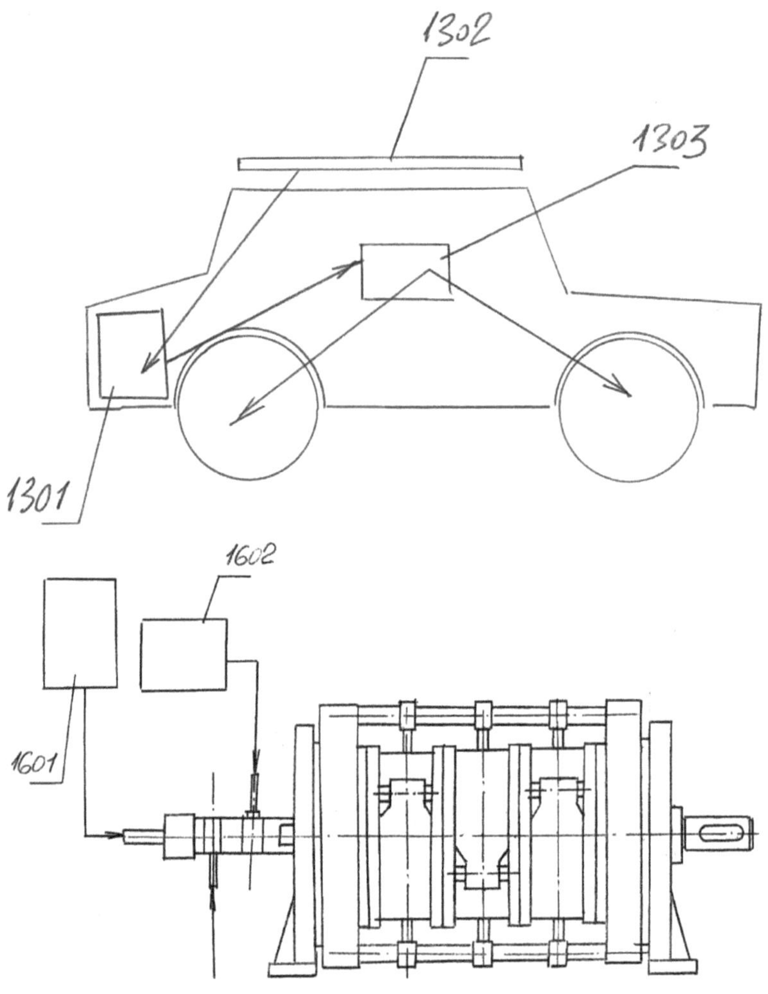



**Figure 14, 15. Schematic diagram of rotary motor in hybrid version.**

**Figure 16, 17. Energy installation in a modern Porsche Taycan electric car.**

As a result of the final geometric shaping, it is possible to obtain an exceptionally high quality of the surface of the structure, without additional mechanical processing and, if necessary, to produce on this surface a coating of a conductive film of artificial diamond on which an electronic component can be attached or soldered.

The proposed composite material after completion of all operations on its manufacture, acquires the form of a finished geometric structure, for example, prisms, which should be considered as a conductive object, in the volume of which dielectric spheres made of synthetic diamonds are uniformly distributed. The cross-section of such a conductor is quite large, and due to the developed volume structure, such a conductor has a low

21

electrical resistance. Since there are inclusions of diamond grains (spheres) in the volume of the electrically conductive structure, which are not a current conductor, the current bypasses these zones in the body of the structure and passes only into the electrically conductive one.

This scheme of current dissipation or distribution over a relatively large cross-section allows to dramatically reduce losses and accelerate current flow. In the case of the need to dissipate heat, the pseudo-porous structure represents the nodes of a specific lattice, in the nodes of which diamond spheres are located, the thermal resistance of which is 4-5 times lower than in the whole structure, so the heat rushes to the nodes of the lattice and this provides a very fast outflow (dissipation) of heat from the source of its origin. That is, in both cases the phenomenon of spotty three-dimensional distribution of zones with different specific coefficients of heat conductivity and electrical conductivity is created.

In addition, the nanometre-scale capsule dimensions and the finish plastic deformation in the cold-fluid regime, allow the gaps between the capsules to be significantly reduced, which improves the efficiency of heat and current pulse extraction and dissipation.

The calculated and expected heat dissipation effect is 4-5 times higher than the best available technical solutions.

Due to the fact that the proposed technical solution affects and can be applied in a number of technological directions in various spheres, the main goal pursued and set in the basic invention is to increase the level of efficiency of the material in terms of heat conductivity and heat dissipation, the rate of heat removal from heating sources and reliability of the process of heat extraction and utilisation during a long period of operation of the object, in which the level of temperature pulsations is stabilised. It is also aimed at increasing the level of material efficiency in terms of electrical conductivity and current dissipation, elimination of current losses during the passage through the structure and reliability of the process of current passage and dissipation during a long period of operation.

The technical solutions that are applied to achieve the objective:

- reducing the capsule diameter to the minimum allowed by the technology of their production (the smaller, the more effective);

- calibration of the geometric shape of the structure by plastic deformation of the capsule shells in the cold-fluid mode; this reduces the volume of voids in the spaces between the capsules, reduces electrical and thermal resistance, improves the mechanical characteristics of the structure and removes internal stresses in the three-dimensional hierarchy of the structure.

Based on the presence of positive effects from the use of composite material, it is possible to assume variants of directions of development and development of the following applications for various fields of application:

**Capsule core** - ceramic; **Capsule shell** - copper; silver; aluminium; nickel;

| Capsule core | Capsule shell |
|---|---|
| - tungsten; | - copper; silver; nickel; aluminium; |
| - iron; | - aluminium; copper; |
| - beryllium; | - aluminium; |
| - magnesium; | - aluminium; |
| - silicon; | - copper; silver; gold ; |
| - zirconium; | - aluminium; |
| - diamond; | - copper; silver; gold; |
| - cital; | - copper; silver; gold; |
| - hard alloy; | - Copper; aluminium; cobalt; molybdenum |

An example of a composite material application:

These composites can be used to manufacture the basis of hard magnetic discs for computer memory drives. Such discs, due to their technical characteristics, have the ability to operate at speeds up to more than 20,000 RPM.

These materials open up new possibilities in:

- creating hybrid discs;
- coating technologies in microelectronics;
- creating activating additives for fuels;

- for the manufacture of critical parts.

As an example of the use of composite material, we can consider the packaging and housing of a semiconductor laser (laser diode), which can then be used in smart home structures.

For an example, consider a laser diode with combined emission and 1 watt optical output power.

A minimum of 1 Ampere of current must be applied to drive the diode and produce 1 watt of output power.

The voltage, taking into account the internal resistance of the laser diode itself and the controlling electronic system, will be at least 2 volts.

Thus, the total power consumption will be 2 watts, with a real output power of 1 watt. The power loss factor of 50% is the best known today.

That is, the least loaded laser diode with this kind of radiation (beam cross section is 300 microns x 1-3 microns) needs to dissipate 1 watt of energy.

The standard housing for this type of diode is designated SOT-148 and its mounting flange diameter is 9 mm.

In order to dissipate such a huge specific amount of heat, a composite material is needed, which is able to dissipate the heat arising from conversion of 1 watt of energy into heat from the hetero-structure of the laser diode, the dimensions of which do not exceed the dimensions of a standard semiconductor crystal of an integrated circuit. The nominal operating temperature in the zone of hetero-structure location cannot exceed +25.,.+27 degrees Celsius. In order to transfer this amount of heat, the hetero-structure is soldered to a composite carrier, which dissipates heat to the diode body, which in turn transfers the resulting heat to the cooling system (thermal electric cooler). The more efficient the material, the more efficient the laser diode performance, including stability, durability and power output.

The problem is much more acute when it is necessary to dissipate heat from a single-beam diode, since in this type of diode the cross-section of the beam is a circle with a diameter of 0.6 microns or less. In this case, the energy concentration is even higher and the heat dissipation and dissipation function becomes even more important.

Considering the fact that just for the needs of all kinds of video systems, optical memory systems, optical memory drives for personal computers and similar products requires a system of laser light sources in different areas of the spectrum, the number of laser diodes, just for these needs is a year more than 100 million pieces, with the price of a laser diode with a power of 1 watt more than $1000.

The bulk of today's laser diodes have an optical power of approximately 80 milliwatts, but operate in the red range of the spectrum, so that the application of a new efficient composite is extremely relevant not only for applications in the smart home and its infrastructure.

As of today, the following composite materials are known to be used for similar purposes:

Copper-tungsten

Copper-molybdenum

Aluminium carbide-silicon aluminium

Aluminium-silicon

Aluminium nitride

Synthetic single-crystal diamond

Chemical diamond

Diamond-Copper Composite. This composite material has the designation DMCH - Diamond-Copper Composite (Diamond Metal Composite for Heat Sink).

The thermal resistance and thermal conductivity of this composite is three times better than that of ordinal composites.

Modern electro-optical systems require much higher performance, 4-5 times better than ordinal composites.

Nanoscale composite material proposed for smart home and smart transport infrastructure applications can provide such results.

The proposed technical solution for use in smart home infrastructure has the following advantages:

- there are only two components in the composite - diamond spheres (grains) and copper shells to them;

- there is a heat-dissipating effect in the composite;

- there is an electric current dissipating effect in the composite;

- the composite has an electrical resistance equivalent to that of copper;

- The composite is formed and calibrated using the real cold-fluid effect of copper (or any other ductile metal);

- The composite has high mechanical strength due to the calibration method of creating a cold-fluid state;

- The composite has a high level of electrical conductivity due to the calibration method of creating a cold-fluid state;

- The composite has more precise dimensions due to calibration by cold drawn of metal or cold metallicity liquid state;

- The composite has a higher level of thermal conductivity due to the very small capsule sizes (nanometres) and due to calibration by creating a cold-fluid state.

Technological processes for the production of diamond nanoscale powders with practically equal granule sizes are absolutely new and have not been used at the enterprises of the world, they can be realised on the basis of institutes of superhard materials.

Technological processes for coating diamond nanoscale powders with copper are also novel.

The use of the above materials allows us to give operating conditions of examples of different supersystems with base units saturated with energy and dependent on dissipation of the resulting heat.

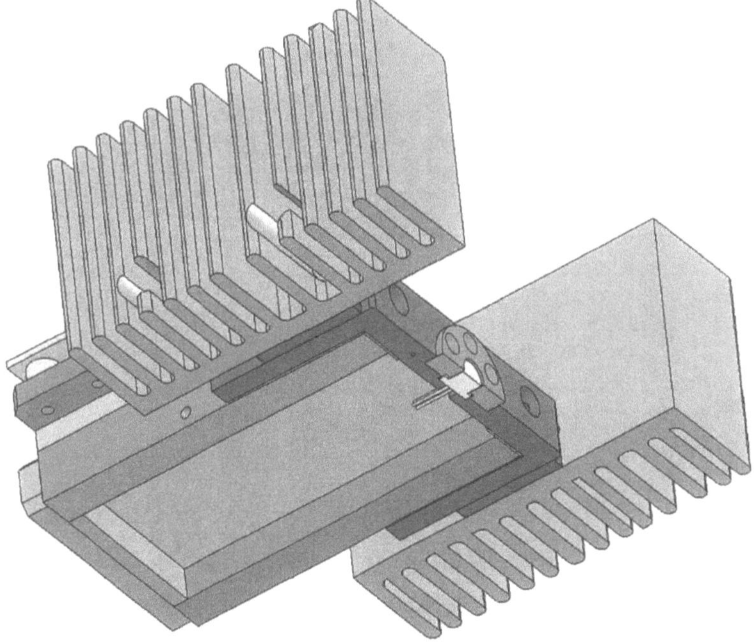

**Figure 18. High-power laser diode module housing in cross-section, in which the housing walls are made of composite material that transfers dissipated heat to the heat sinks through thermoelectric cooling elements.**

**Figure 19. Components in real execution**

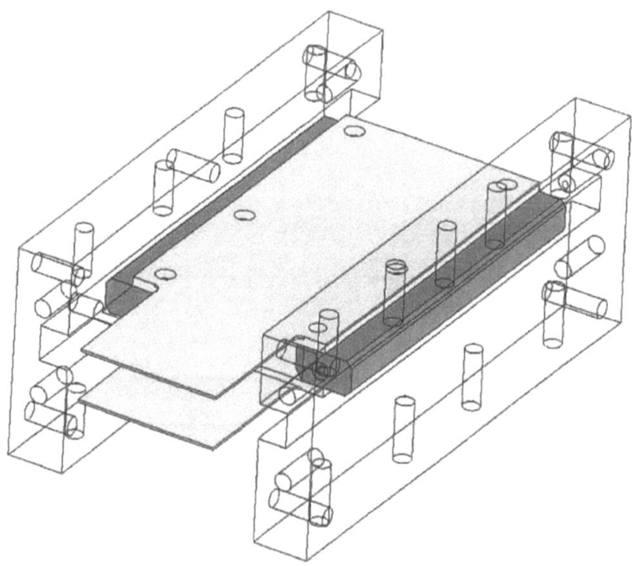

**Figure 20. Three-dimensional model of the laser diode module, in which electronic printed circuit boards (yellow in the figure) are fixed on a board (red) made of composite material and having high possibilities of heat dissipation and permanent cooling of printed circuit boards.**

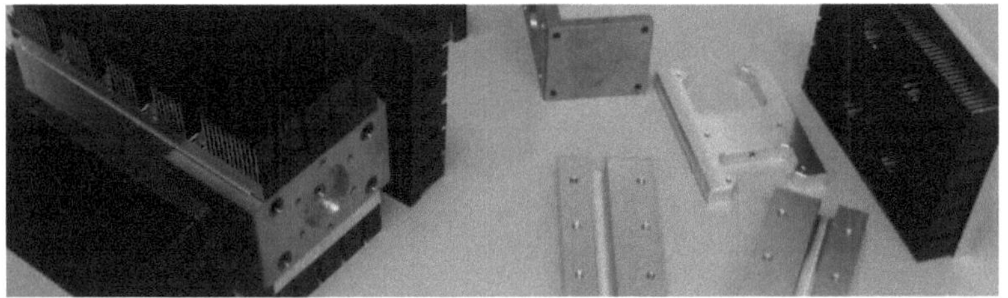

**Figure 21. The same elements in actual composite fabrication with copper and aluminium capsule shells.**

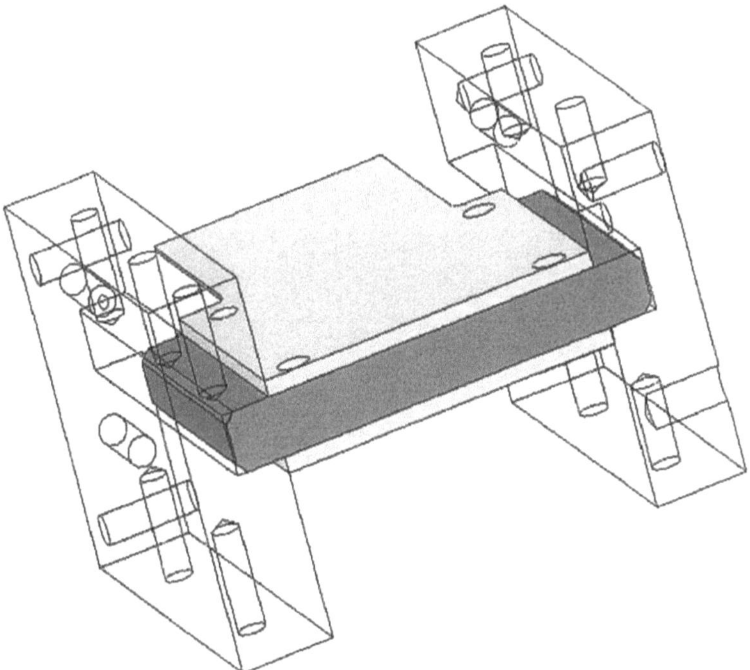

**Figure 22: Three-dimensional model of a laser diode module in which electronic printed circuit boards (yellow in the figure) are fixed on a board (red) made of composite material and having high heat dissipation capabilities and constant cooling of the printed circuit boards.**

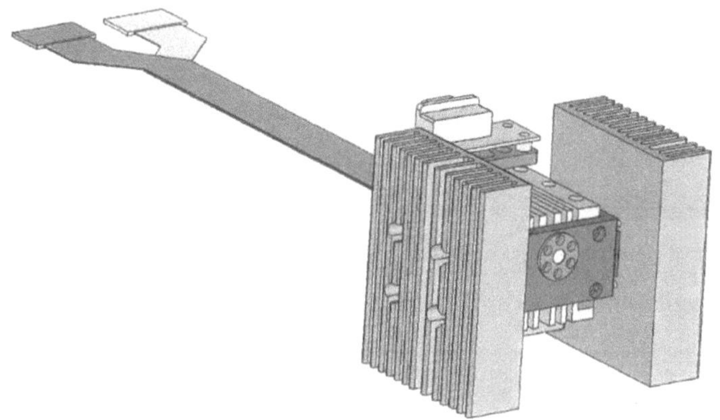

**Figure 23. High-power laser diode module with main parts made of composite material**

**Figure 24. Novel instrumentation (mobile phone-sized) for online real-time control of high-power laser diode emission parameters**

The emergence of such devices allows for full control and analytical evaluation of the elements of the smart home infrastructure that require significant energy inputs for operation and control within the structure of smart home supersystems and subsystems.

At a relatively low cost and simplicity, the effect of using these devices is extremely high and allows the control and optimisation of the smart home infrastructure elements using solar panels and the gradual accumulation of excess solar energy from the daytime in storage components with cooling systems made of composite material.

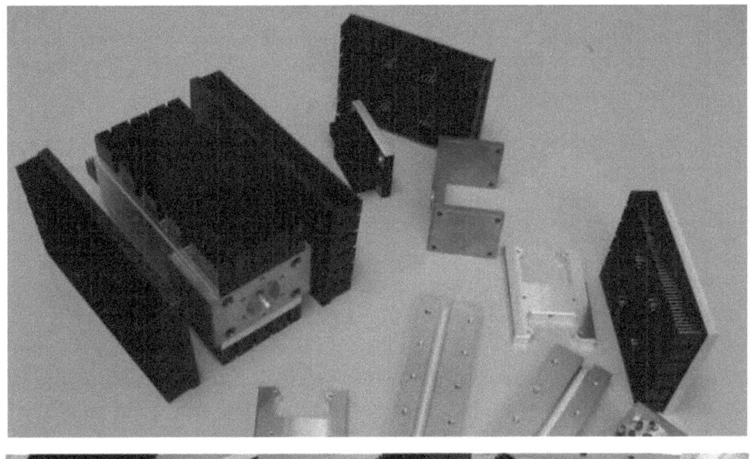

**Figure 25, 26. Details of laser diode module housing devices**

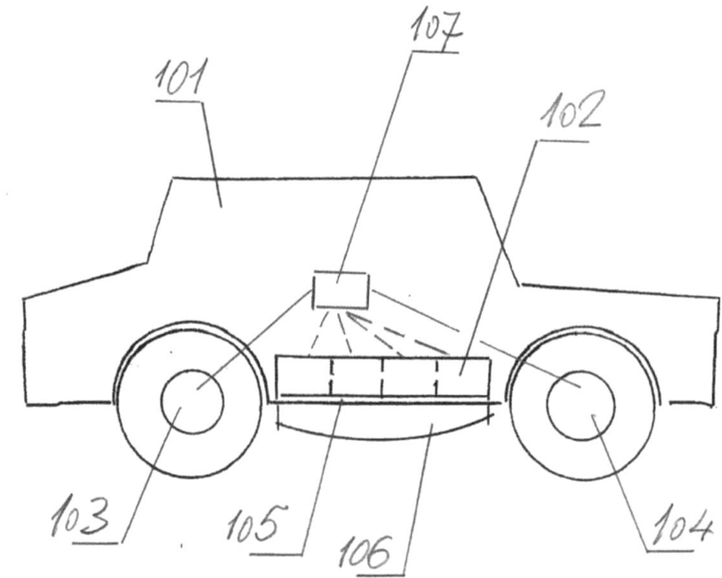


**Figure 27. Batteries of an electric vehicle that undergoes recharging in a smart home environment and whose enclosures are made of composite material.**

The radiators of these batteries are also designed to utilise the heat dissipation potential provided by the composite material from which the air contacting elements of the battery housings and baffles are made.

The structure of the composite material in the form of a pseudo-porous volume significantly facilitates online monitoring and control of energy management processes with the additional use of sensor modules based on the principles of electromagnetic resonance spectroscopy.

Further development of system links between smart home and smart transport elements using electromagnetic resonance spectroscopy on monitoring and control lines facilitates the application of artificial intelligence elements and artificial neural networks.

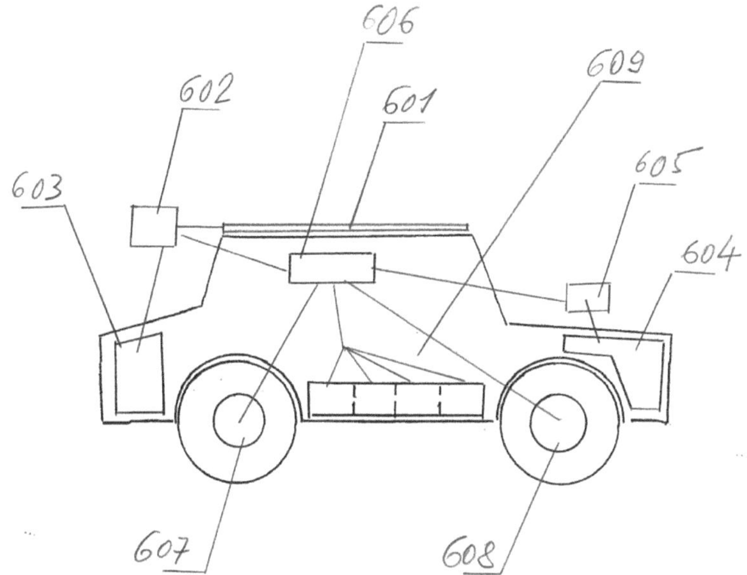


**Figure 28. System development of electric vehicle energy supply elements (and in the future elements of smart home infrastructure) with the inclusion of a solar panel (position - 601) and two control processors (position 602 and 605) regulating energy storage with subsequent recharging of the main battery (group position 609).**

The presence of an active cooling system due to the properties of the composite material allows the structure shown in the figure to significantly reduce the losses that occur in battery and accumulator systems in operation today.

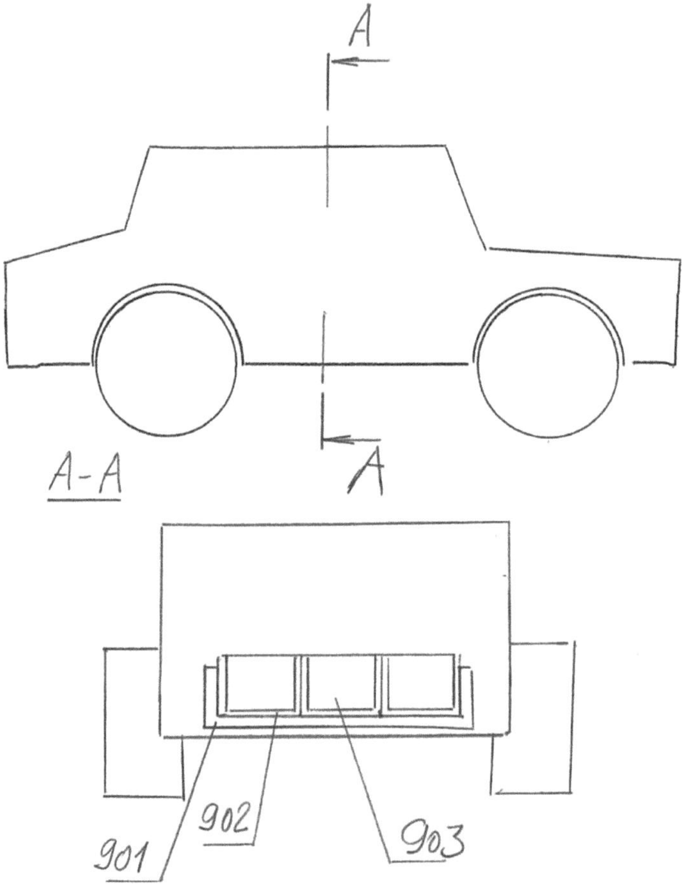



**Figure 29. The possibility of obtaining a battery or battery system with 12 batteries or more in a battery system due to the fact that the composite material successfully dissipates the generated heat, which dramatically reduces the thermal and destructive stresses on the batteries (item 903) and their housing parts (items 901 and 902).**

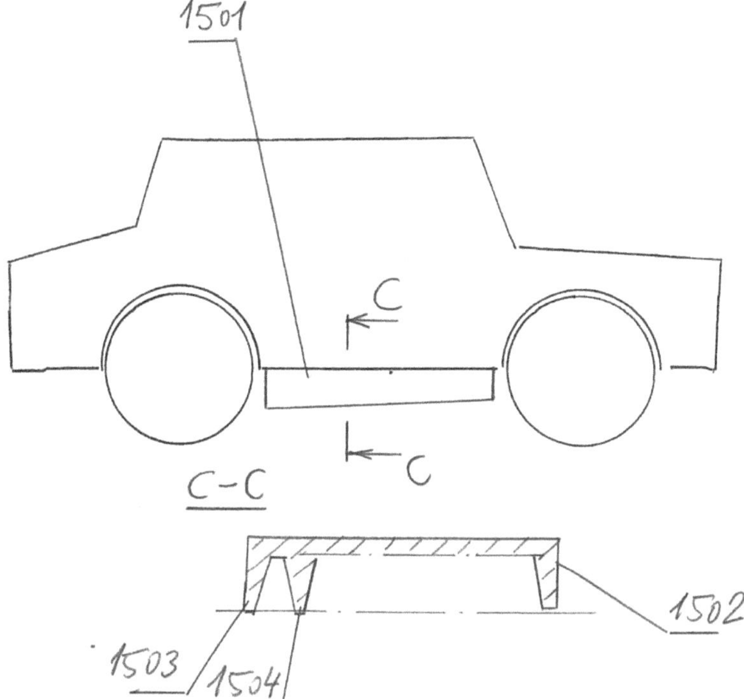


**Figure 30. The system arrangement shown in the figure allows the most efficient radiators (item 1501) to be used with the most workable radiator tooth profiles (items 1503 and 1504).**

Similar concepts are applicable with the same effect in smart home and smart transport systems and their infrastructure elements.

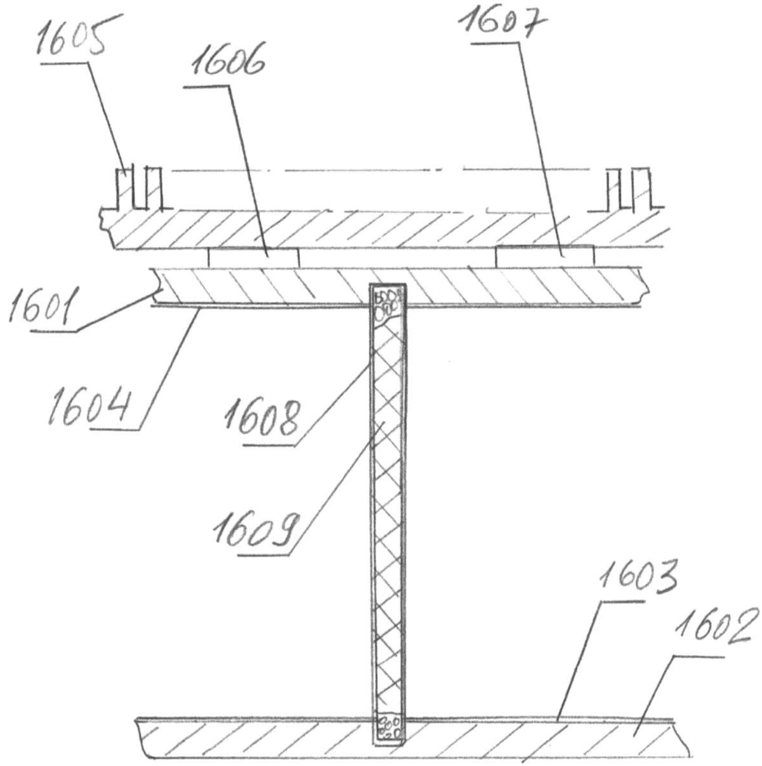



**Figure 31. Structure of the battery case (1601) with composite connections (1608 and 1609) transferring heat through thermoelectric coolers (1606 and 1607) to the heat sink (1605).**

The system is as simplified as possible, and the heat dissipation rate is significantly higher than in systems adopted today.

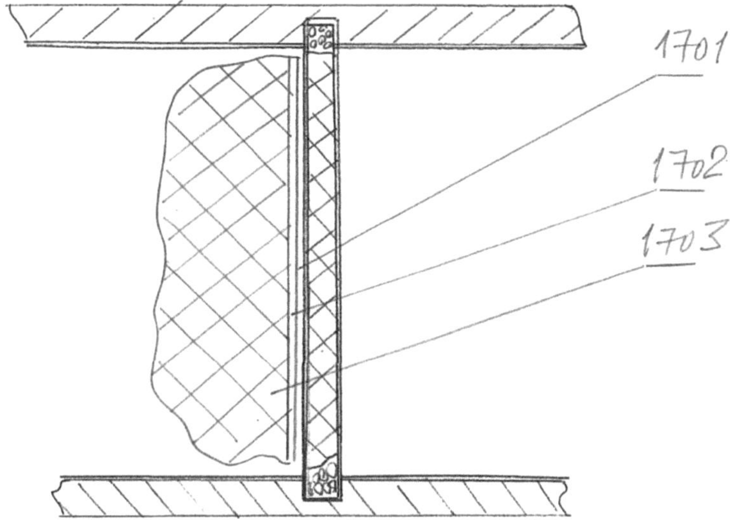


**Figure 32. Further explanations of the system of relationships between the structural elements of batteries and accumulators utilising the proposed composite material and its heat dissipation properties.**

In addition, the batteries show the possibility of using also carbon composites as electrodes in the form of carbon wool (1703) in a carbon cloth shell (1701 and 1702).

As this type of composite material is a conductor of electric current, the elastic format of the electrodes allows for maximum contact and, due to this, with a high heat dissipation coefficient, the elasticity of the electrodes and their sheaths ensure maximum performance of the system as a whole.

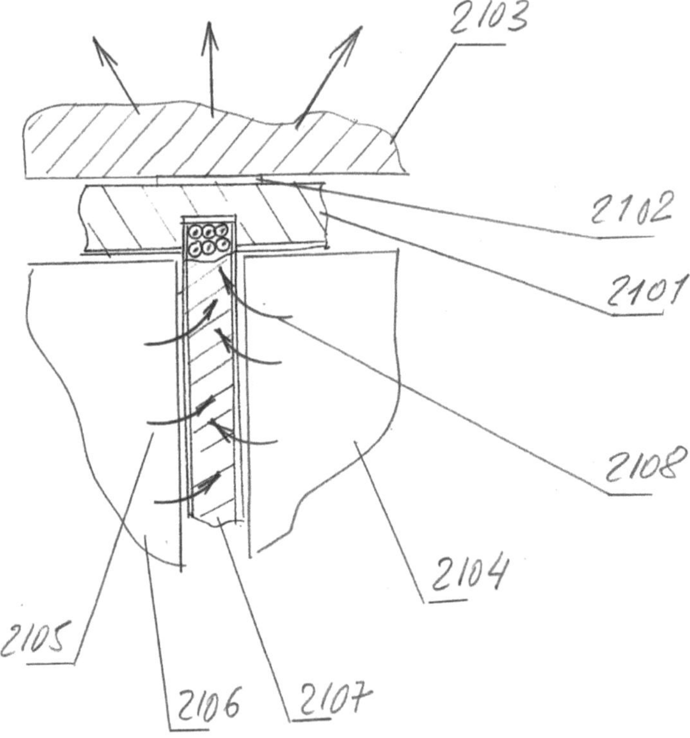


**Figure 33. Trajectories of heat dissipation heat transfer from the elastic electrodes (2106 and 2104) to the composite partition (2107), transferring it to the module housing wall (2101), which in turn, through the thermoelectric cooler (2102), transfers heat to the radiator (2103).**

This scheme is very easy to control and provides a high level of efficiency.

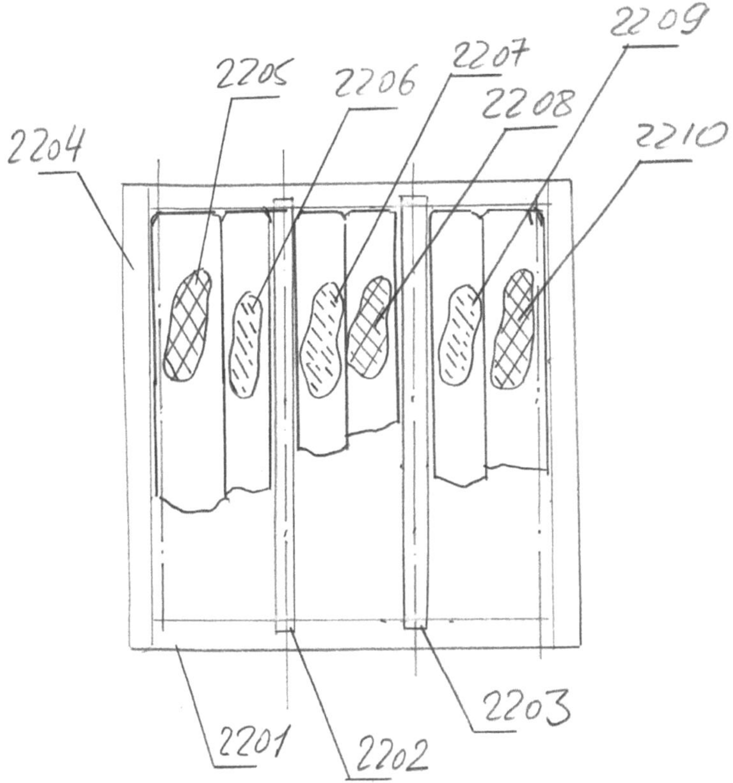



**Figure 34. Schematic of the formation of a battery or accumulator with electrodes made of carbon composite wool (2206, 2207, 2209) with carbon composite fabric shells (2205, 2208, 2210), wherein the battery or accumulator body is made of the proposed composite material (2204, 2201, 2202, 2203).**

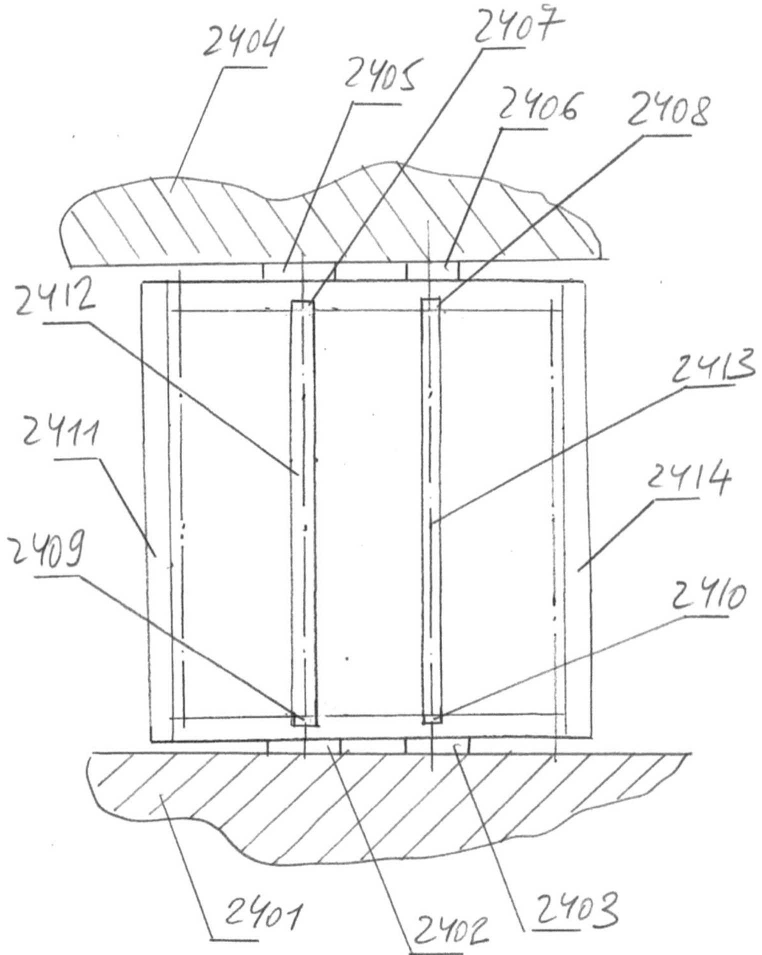


Figure 35. Schematic diagram of a battery or accumulator with conventional electrodes in a housing made of the proposed composite material (parts 2409, 2410, 2411, 2412, 2413, 2414), with contact with the surrounding structural elements by means of thermoelectric coolers (2405, 2406, 2402, 2403).

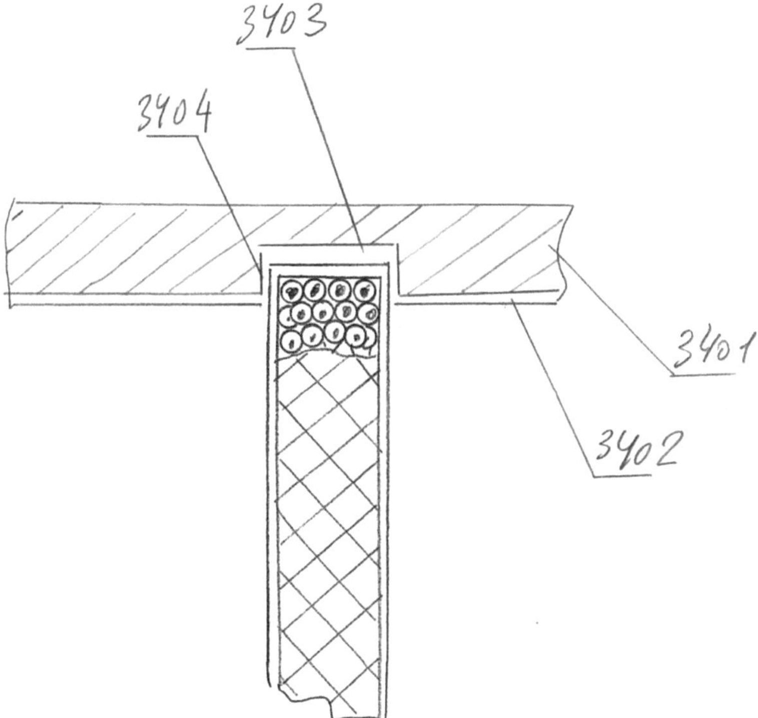


**Figure 36. Nature of interaction between structural elements of battery or battery cases with structural elements made of the proposed composite material.**

Composite materials can be represented as:

- conductive and dissipating pulses of electric current;

- heat-conductive and heat-dissipative;

- conductive and heat-dissipative;

- heat conducting and dissipating pulses of electric current.

Capsule composite compounds based on microcomponents in the form of spherical multilayer capsules having a size factor in the nanoscale metric range.

41

The proposed capsule composite material is a three-dimensional composite pseudo spongy structure that includes a plurality of multilayered, interconnected components having an equivalent (identical) geometric shape and each comprising at least two plies.

The structure consists of a plurality of components in contact with each other and forming a complete three-dimensional geometric shape, in which said components are uniformly and equivalently distributed over the volume and have and form at all points of the formed complete three-dimensional geometric shape equal conditions of electrical and thermal interaction with each other, wherein the same-type layers of all components are separated from each other by the same-type layers of the same components.

The three-dimensional structure of the composite material is characterised by the fact that each layer of each component is a closed three-dimensional geometric figure.

The three-dimensional structure of the composite material, characterised in that each successive layer in each component covers the entire surface of the previous layer in each component.

Thus, structurally, the proposed invention can be represented as an integrative hierarchy consisting of interrelated distinctive physical, constructive and technological features, on the basis of which the final properties of the composite material are formed.

The material has thermal conductive and electrically conductive properties at the same time. The material has the buffering ability to dissipate in its volume thermal pulses and associated electric current pulsations.

The material is capable of fundamentally changing the operating conditions and performance characteristics of energy-rich electronic devices; it makes it possible to create a new generation of electronic devices that are much less dependent on thermal characteristics. This is especially important for high-power pulse technology, which has a power at the peak of the pulse greater than the nominal power of the device.

## Distinctive features of the composite material

Final version of the distinctive features of a composite material for use as a structural material in infrastructural elements, subsystems and supersystems of a smart home:

A composite material formed from a plurality of multilayered three-dimensional capsules, each comprising at least two layers, and bonded together along the outer surfaces of their plastically deformed outer layers;

The composite material according to claim 1, characterised in that it comprises a plurality of multilayered three-dimensional capsules each comprising a core of spherical shape and at least one shell repeating on its inner surface the geometric shape of the core;

The composite material according to claim 1, characterised in that different structural materials are used for the core and the shells;

The composite material according to claim 1, characterised in that the core and shell structural materials differ in physical and chemical properties;

The composite material according to claim 1, characterised in that the core and shell structural materials differ in mechanical properties;

The composite material according to claim 1, characterised in that the core and shell structural materials differ in hardness;

The composite material according to claim 1, characterised in that the core and shell structural materials differ in ductility;

The composite material according to claim 1, characterised in that the core and shell structural materials differ in electrical conductivity;

The composite material according to claim 1, characterised in that the core and shell structural materials differ in thermal conductivity;

The composite material according to claim 1, characterised in that the hardness of the core structural material is higher than the hardness of the shell structural material;

The composite material according to claim 1, characterised in that the core and shell structural materials differ in thermal conductivity, wherein the thermal conductivity of the core structural material is at least 2 times higher than the thermal conductivity of the shell structural material;

The composite material according to claim 1, characterised in that the hardness of the core structural material is higher than the hardness of the shell structural material, the shells being generally made of ductile metal;

The composite material according to claim 1, characterised in that the hardness of the core structural material is higher than the hardness of the shell structural material, the shells being generally made of a ductile metal and the core of a non-metallic dielectric material;

The composite material according to claim 1, characterised in that, where the number of shells is greater than one, the hardness of the structural material of each preceding shell is greater than that of the succeeding shell;

The composite material according to claim 1, characterised in that, where the number of shells is greater than one, the hardness of the structural material of each preceding shell is greater than that of the succeeding shell, and the ductility of the structural material of each succeeding shell is greater than that of the preceding shell;

The composite material according to claim 1, characterised in that diamond is used as the core structural material;

The composite material according to claim 1, characterised in that copper is used as the structural material of the unit shell.

## Autonomous energy-producing plant

For a smart home or smart transport, there is a need for a reliable, inexpensive and easy to manage independent source of backup energy.

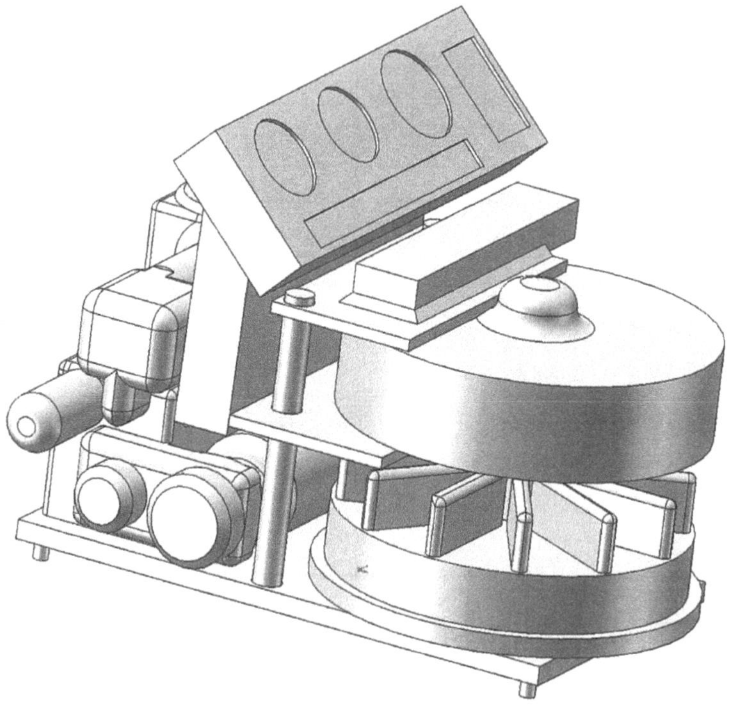

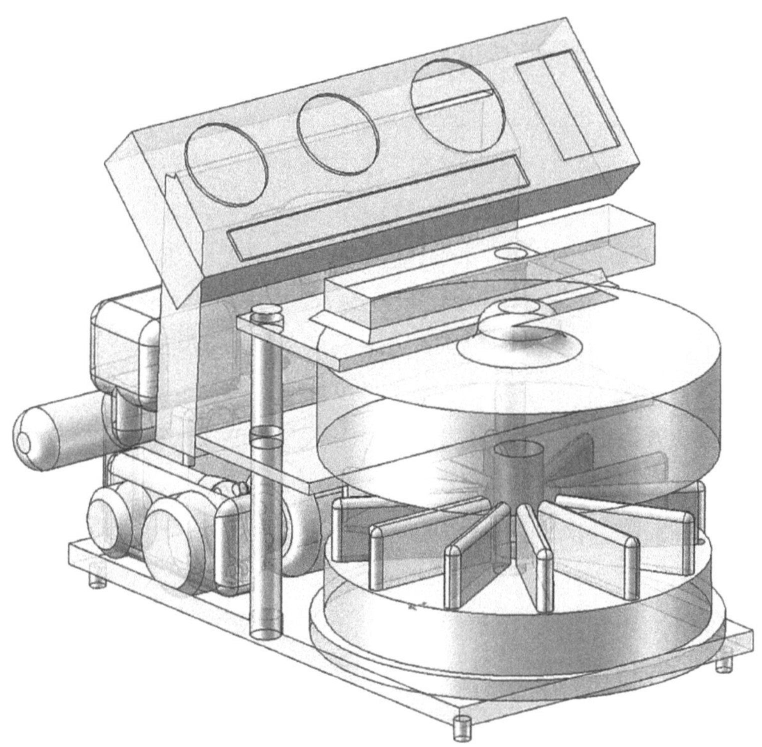

**Figure 37, 38. Autonomous energy generating unit of modular type for installation in smart home or smart transport infrastructure as a backup energy source**.

The plant built on the basis of integrative technical solutions aimed at saving fossil fuel in internal combustion engines; at highly efficient conversion of energy types with increasing output torque, and, the use of planar technology in the design of low-speed electric generators.

The proposed plant is based on the principle of successive stepwise conversion of energy types, with gradual amplification of equivalent power characteristics at the output of each conversion step.

The first conversion step is the conversion of the combustion energy of the fossil fuel into mechanical rotational energy of the output shaft of the internal combustion engine used as an energy converter in the first step. The energy gain is the use of a fuel composite comprising a hydrodynamically mixed and aerodynamically foamed fuel composition of ethanol and water, in a proportion of 70% ethanol and 30% synthetic water derived from air with impurities of ethanol vapour or other alcohol or gas condensate. Deeply purified water may also be used;

The second stage of conversion is the conversion of mechanical energy of rotation of the output shaft of the internal combustion engine into electrical energy with certain parameters;

The third stage of conversion is generation of electric current pulses of certain parameters;

The fourth most important conversion step is the conversion of the current pulses into torque at the output shaft of the torque amplifier converter. The amplification effect is achieved due to two factors: the first factor is kinematic, and it consists in the use of a rotary converter of linear motion into rotary motion, which eliminates dead spots and energy losses associated with overcoming them. The second factor is electromagnetic, and it consists in the use of armoured electromagnets with a special design of the magnetic core and a special design of the solenoid, with additional constructive advantages of the combination of the core and planar design of the solenoid coil;

The fifth conversion step is to apply torque to the shaft of the planar low-speed generator and generate electricity, which is the output product of the plant;

All these advantages determine the efficiency of electromagnets belonging to the category of so-called warm or cold magnets and having a pulling or pushing force that is an order of magnitude higher than the equivalent power consumption for its generation. The use of a planar low-speed generator makes it possible to use the advantages of the power characteristics of electromagnets in their most efficient mode of operation, since the lower the frequency of operation of the electromagnet, the longer the duration of the working cycle of the electromagnet and the more time is used to restore the magnetic properties of the combined magnetic core.

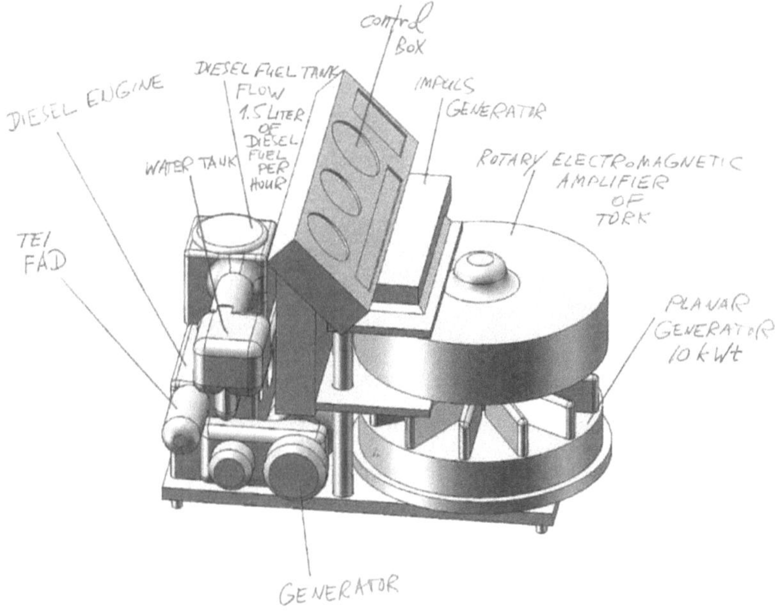

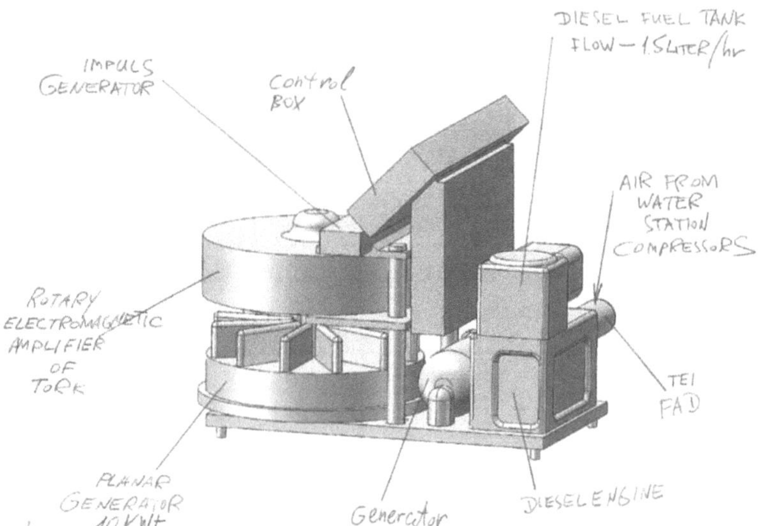


**Figure 39, 40. Autonomous energy generating unit of modular type for installation in smart home or smart transport infrastructure as a backup energy source with description of the components in the composition.**

All principal technical solutions underlying the proposed installation have substantial novelty and are inventions. These include:

- use as a combustible mixture of an internal combustion engine - a mixture of an organic base with an inorganic base;

- application of hydrodynamic activators working on the Bernoulli principle for mixing of organic and inorganic components of the combustible mixture, carried out in a constantly moving flow of the organic component;

- application of aerodynamic activators working on the Bernoulli principle for foaming of composite - compound combustible mixture, carried out in a constantly moving flow of combustible mixture.

Thus, the use of the latest composite materials and backup energy source installations leads to an increase in derivative efficiency and power output. At the same time, a reduction of temperatures inside the technical

50

mechanisms and in the outdoor environment is achieved. It also achieves a reduction in the size of the energy units and a more rational use of space in the server rooms of the smart home and the engine compartments of smart vehicles.

## New lithium-ion battery design

To form a comprehensive technical solution for the design and duty cycle of lithium-ion batteries to build the concept of this innovation in the book, the following innovations are available to automotive manufacturers today:

- technology and design of volumetrically porous electrodes made of (composite) carbon wool;

- Technology of high-speed electrochemical coatings of lithium on (composite) carbon wool;

- technology of forming conductive and water-permeable covers and contacts from carbon (composite) fabric for battery electrodes;

- technology and design for forming diamond and copper composite panels (pseudoporous electrical current dissipating and heat dissipating materials);

- technology of using neutral membranes from polypropylene fabric in electrochemical cells.

By applying the above technologies, the battery gains the following advantages over known batteries of the same type:

- dramatic increase in efficiency due to the increased contact area of electrodes with electrolyte (more than 1000 times);

- significant temperature reduction in the working volume of the battery;

- Increase the active battery life;

- absolute environmental friendliness of battery disposal;

- reducing the amount of lithium for the battery electrodes;

- reducing the production cost of battery cells;

- Increasing the energy yield of battery operation;

- increasing the electrical capacity of the battery;

- increasing the level of safety in battery operation.

# List of references used, patent and licence information:

APPENDIX 1-1

**United States Patent Application** | **20210104744**

**Kind Code** | **A1**

**OGUNI; Teppei; et al.** | **April 8, 2021**

---

SECONDARY BATTERY AND MANUFACTURING METHOD THEREOF

## Abstract


Provided is a layer for preventing a short circuit between a positive electrode and a negative electrode in a solid *battery* using a layer containing a solid electrolyte. As the solid electrolyte between the positive electrode and the negative electrode, a layer containing a graphene compound is used. Lithium ions can pass through the layer containing the graphene compound. Lithium ions are added in advance in the layer containing the graphene compound. Specifically, a modifier is used, and a graphene compound chemically modified with a functional group such as ether and ester with an increased interlayer distance is used.

APPENDIX 1-2

**United States Patent Application**  20210104719

**Kind Code**  **A1**

**OIKAWA; Makiko**  **April 8, 2021**

BATTERY ELECTRODE AND METHOD FOR MANUFACTURING THE SAME

## Abstract

There is provided a method for manufacturing a *battery* electrode. The method includes: forming a precursor of the *battery* electrode including a double-sided coating area in which both sides of a current collector are coated with an electrode material layer and a single-sided coating area adjacent to the double-sided coating area; subjecting the current collector located at a boundary portion between the double-sided coating area and the single-sided coating area to a heat treatment locally; and pressurizing the precursor of the *battery* electrode. The single-sided coating area includes a main side of the current collector that is coated with the electrode material layer.

**United States Patent Application**                    20210100579

**Kind Code**                                                            **A1**

**Shelton, IV; Frederick E.; et al.**                    **April 8, 2021**

---

MODULAR BATTERY POWERED HANDHELD SURGICAL
INSTRUMENTS AND METHODS THEREFOR

### Abstract


Disclosed is a method of controlling a modular *battery* powered handheld surgical instrument. The surgical instrument including a *battery,* a user input sensor, a controller, a radio frequency (RF) drive circuit, an ultrasonic transducer, an ultrasonic transducer drive circuit, and an end effector. The end effector including an electrode electrically coupled to the RF drive circuit, an ultrasonic blade acoustically coupled to the ultrasonic transducer, and a sensor to measure tissue parameters. The method includes applying an RF current drive signal to the electrode by the RF drive circuit; applying an ultrasonic drive signal to the ultrasonic transducer by the ultrasonic transducer drive circuit to acoustically excite the ultrasonic blade; controlling intensity, wave shape, and/or frequency of the RF current drive signal and the ultrasonic drive signal on a sensed measure of a tissue or user parameter.

**United States Patent Application** 20210091416

**Kind Code** A1

**SEKI; Hayato; et al.** **March 25, 2021**

---

SECONDARY BATTERY, BATTERY PACK, VEHICLE, AND STATIONARY POWER SUPPLY

## Abstract

A secondary *battery* includes a positive electrode, a first aqueous electrolyte held on the positive electrode, a negative electrode, a second aqueous electrolyte held on the negative electrode, and a separator interposed between the positive electrode and the negative electrode. A difference between an osmotic pressure (N/m.sup.2) of the first aqueous electrolyte and an osmotic pressure (N/m.sup.2) of the second aqueous electrolyte is 90% or less (including 0%) of the higher one of the osmotic pressure of the first aqueous electrolyte and the osmotic pressure of the second aqueous electrolyte.

**United States Patent Application**                    20210091402

**Kind Code**                    **A1**

**Londarenko; Yuriy Y.**                    **March 25, 2021**

---

MULTI-LAYER BATTERY CONFIGURATIONS

## Abstract


Rechargeable *battery* cells according to embodiments of the present technology may include a *housing* including a first conductive segment operable at anode potential, and a second conductive segment operable at cathode potential. The *housing* may include a gasket positioned between the first conductive segment and the second conductive segment and configured to hermetically seal the *housing*. The *battery* cells may also include an electrode stack. The electrode stack may include a cathode current collector having a cathode active material extending across a first surface of the cathode current collector. The cathode current collector may be characterised by at least two pleats. The electrode stack may also include an anode current collector having an anode active material extending across a first surface of the anode current collector. The anode current collector may be characterised by at least two pleats.

APPENDIX 1-6

**United States Patent Application**                    **20210091363**

**Kind Code**                                                            **A1**

**Lane; Robert Clinton**                             **March 25, 2021**

---

High Voltage Battery Module Parallel Cell Fusing System

## Abstract

A fUsing system for a brick of lithium-ion *battery in a battery* module is provided where the fusing system has a combination of low-voltage fuses and a high-voltage fuse. The low-voltage fuse can have one or more fusing elements in a springy spiral configuration or a straight configuration with the fuse element encapsulated.

APPENDIX 1-7

**United States Patent Application**                    **20210091360**

**Kind Code**                                                            **A1**

**Balaram; Haran; et al.**                            **March 25, 2021**

---

BATTERY PACK WITH ATTACHED SYSTEM MODULE

## Abstract

*Battery* systems according to embodiments of the present technology may include a *battery*. The *battery* may include a first electrode terminal and a second electrode terminal accessible along a first surface of the *battery*. The systems may include a module electrically coupled with the *battery*. The module may include a circuit board characterised by a first surface and a second surface opposite the first surface. The module may include a mould extending from the first surface of the circuit board towards the *battery*. The module may include a first conductive tab electrically coupling the module with the first electrode terminal. The module may include a second conductive tab electrically coupling the module with the second electrode terminal. The second conductive tab may extend across the mould substantially parallel to the first surface of the circuit board.


APPENDIX 1-8

**United States Patent Application**          20210075063

**Kind Code**          A1

**Dou; Shushi; et al.**          **March 11, 2021**

---

LITHIUM-ION BATTERY AND APPARATUS

**Abstract**

This application provides a lithium-ion *battery* and an apparatus. The lithium-ion *battery* includes an electrode assembly and an electrolyte. The electrode assembly includes a positive electrode plate, a negative electrode plate, and a separator. A positive active material of the positive electrode plate includes Li.sub.x1Co.sub.y1M.sub.1- y1O.sub.2-z1Q.sub.z1, where 0.5.ltoreq.x1.ltoreq.1.2, 0.8.ltoreq.y1.ltoreq.1.0, 0.ltoreq.z1.ltoreq.0.1, M is selected from one or more of Al, Ti, Zr, Y, and Mg, and Q is selected from one or more of F, Cl, and S. The electrolyte contains an additive A, an additive B, and an additive C. The additive A is a polynitrile six-membered nitrogenheterocyclic compound with a relatively low oxidation potential. The additive B is a silyl phosphite compound or a silyl phosphate compound or a mixture thereof. The additive C is a halogen substituted cyclic carbonate compound.

**United States Patent Application**                     20210075060

**Kind Code**                                           **A1**

**NAKAYAMA; Tetsuri**                                   **March 11, 2021**

---

NON-AQUEOUS ELECTROLYTE SECONDARY BATTERY

### Abstract

A non-aqueous electrolyte secondary *battery* disclosed herein includes a positive electrode, a negative electrode, and a non-aqueous electrolyte. The positive electrode includes a positive electrode current collector and a positive electrode active material layer provided on the positive electrode current collector. The non-aqueous electrolyte contains lithium fluoro - sulfonate. The positive electrode active material layer contains a positive electrode active material. The positive electrode active material layer contains hydrated alumina at least in a surface layer portion.

APPENDIX 1-10

**United States Patent Application**                     20210075015

**Kind Code**                                           **A1**

**LEE; Jungmin; et al.**                                **March 11, 2021**

---

SECONDARY LITHIUM BATTERY ANODE AND SECONDARY LITHIUM BATTERY INCLUDING THE SAME

### Abstract

Disclosed are a secondary lithium *battery* anode and a secondary lithium *battery* including the same, the secondary lithium *battery* anode comprising a current collector and an anode active material layer located on at least one surface of the current collector, wherein the anode active material layer includes an anode active material layer which has sphericity of 0.83 to 0.91, and a binder which has an average particle diameter (D50) of 180 nm to 450 nm.

ANNEX 1-11

**United States Patent Application**                    20210075008

**Kind Code**                                                              **A1**

**Park; Benjamin Yong; et al.**                       **March 11, 2021**

---

Prelithiated And Methods for Prelithiating an Energy Storage Device

## Abstract

The present disclosure relates to prelithiated Si electrodes, methods of prelithiating Si electrodes, and use of prelithiated electrodes in electrochemical devices are described. There are several characteristics of electrode prelithiaiton that enable the superior *battery* performance. First, a prelithiated silicon anode is already in its expanded state during SEI formation, and therefore less of the SEI layer breaks down and reforms during cycling. Second, the prelithiated anode has a lower anode potential, which may also help the cycle performance of an electrochemical device.

ANNEX 1-12

**United States Patent Application**          **20210066712**

**Kind Code**                                              **A1**

**KIM; Young-Ki; et al.**                        **March 4, 2021**

---

POSITIVE ELECTRODE ACTIVE MATERIAL FOR RECHARGEABLE LITHIUM BATTERY AND RECHARGEABLE LITHIUM BATTERY INCLUDING SAME

### Abstract

The present invention relates to a positive electrode active material for a rechargeable lithium *battery* and a rechargeable lithium *battery* including same, wherein the positive electrode active material comprises a core and a surface layer formed on the surface of the core, the core comprising a first crystalline structure, the surface layer comprising a first crystalline structure and a second crystalline structure different from the first crystalline structure, the first crystalline structure being present more than the second crystalline structure in the surface layer.

ANNEX 1-13

**United States Patent Application**          **20210066683**

**Kind Code**                                              **A1**

**Lane; Robert Clinton**                        **March 4, 2021**

---

Method to Prevent or Minimize Thermal Runaway Events in Lithium Ion Batteries

### Abstract

A method to prevent or minimise an occurrence of a thermal runaway event in a *battery* module of an electric vehicle. The method places a gas barrier between a venting space and a wall of each *battery* cell so that escaped gas from one *battery* cell does not impinge onto another *battery* cell.

ANNEX 1-14

**United States Patent Application** 20210057777

**Kind Code** A1

**SUGIYO; Takeshi ; et al.** **February 25, 2021**

---

ALL-SOLID-STATE BATTERY, METHOD FOR MANUFACTURING THE SAME, AND PROCESSING DEVICE

## Abstract

The present invention prevents edge collapse of an electrode layer of a laminated body included in an all-solid-state *battery*. A method of producing an all-solid-state *battery* includes: a laminated body forming step of forming a laminated body (310) including (i) a positive electrode layer (302), (ii) a negative electrode layer (304) having a polarity opposite of a polarity of the positive electrode layer (302); and (iii) a solidelectrolyte layer (303) disposed between the positive electrode layer (302) and the negative electrode layer (304); and a cutoff step of cutting off an outer peripheral edge of the laminated body (310) so as to form a laminated body containing a powder material.

**United States Patent Application**                   **20210057755**

**Kind Code**                                                 **A1**

**Brewer; John C. ; et al.**                   **February 25, 2021**

---

## ANODES FOR LITHIUM-BASED ENERGY STORAGE DEVICES

### Abstract

An anode for a lithium-based energy storage device such as a lithium-ion *battery* is disclosed. The anode includes a current collector having an electrically conductive layer and a surface layer overlaying the electrically conductive layer. A lithium storage layer is overlaying the surface layer and the surface layer includes a metal chalcogenide having at least one of sulfur or selenium. The metal chalcogenide may include a metal sulfide, a metal polysulfide, a metal selenide, a metal polyselenide, or a combination thereof. The metal chalcogenide may include a copper sulfide or a copper polysulfide. The lithium storage may include a total content of silicon, germanium, or a combination thereof of at least 40 atomic %. The lithium storage layer may be a continuous porous lithium storage layer having an average density from about 1.1 g/cm.sup.3 to about 2.25 g/cm.sup.3 and comprises at least 85 atomic % amorphous silicon.

**United States Patent Application**                    20210057721

**Kind Code**                                                           **A1**

**KAWASAKI; Daisuke ; et al.**                    **February 25, 2021**

---

LITHIUM-ION SECONDARY BATTERY


**Abstract**

Provided is a lithium ion secondary *battery* having high energy density and excellent cycle characteristics, and hardly causing burning. The present invention relates to a lithium ion secondary *battery* comprising an electrode mixture layer comprising an electrode active material comprising Si alloy having a median diameter of 1.2 .mu.m or less and 12% by weight or more and 50% by weight or less of an electrode binder; and an electrolyte solution comprising 60% by volume or more and 99% by volume or less of a phosphoric acid ester compound, 0% by volume or more and 30% by volume or less of a fluorinated ether compound, and 1% by volume or more and 35% by volume or less of a fluorinated carbonate compound, wherein the total amount of the phosphoric acid ester compound and the fluorinated ether compound is 65% by volume or more.

**United States Patent Application**            20210057690

**Kind Code**                                    **A1**

**Fukutome; Kazuaki ; et al.**           **February 25, 2021**

---

## BATTERY PACK AND METHOD FOR PRODUCING THE SAME

### Abstract

A *battery* pack includes a plurality of secondary *battery* cells, a *battery* holder, a circuit substrate, and an outer case, in which the *battery* holder is divided into a plurality of divided holders, each of the divided holders forms a fitting structure for fitting the divided holders to each other at an interface for joining the divided holders together, the *battery* holder forms a substrate holding area that holds a circuit substrate with the circuit substrate surrounded by side walls in a state where the divided holders are coupled to each other by the fitting structure, a joint interface where the divided holders are fitted to each other by the fitting structure of the divided holders is exposed in the substrate holding area, and a surface of the circuit substrate is covered with a potting resin in the substrate holding area.



**United States Patent Application**            20210057687

**Kind Code**                                    **A1**

**MATSUO; Tatsumi ; et al.**             **February 25, 2021**

---

## BATTERY DEVICE AND MANUFACTURING METHOD

### Abstract

A *battery* device includes a *battery* cell; an exterior member that accommodates the *battery* cell; and one or more foamed resin fixing members disposed between the *battery* cell and the exterior member. The foamed resin fixing members are formed of a foamed resin having self-adhesiveness.


ANNEX 1-19

**United States Patent Application**                           20210053689

**Kind Code**                                                          **A1**

**Lynn; Robert; et al.**                              **February 25, 2021**

---

VEHICLE CABIN THERMAL MANAGEMENT SYSTEM AND METHOD

### Abstract


The system can include an on-board thermal management subsystem. The system 100 can optionally include an off-board (extravehicular) infrastructure subsystem. The onboard thermal management subsystem can include: a *battery* pack, one or more fluid loops, and an air manifold. The system 100 can additionally or alternatively include any other suitable components.

**United States Patent Application** 20210052313

**Kind Code** **A1**

**Shelton, IV; Frederick E. ; et al.** **February 25, 2021**

---

# MODULAR BATTERY POWERED HANDHELD SURGICAL INSTRUMENT WITH SELECTIVE APPLICATION OF ENERGY BASED ON TISSUE CHARACTERISATION

## Abstract

A surgical instrument comprises a shaft assembly comprising a shaft and an end effector coupled to a distal end of the shaft; a handle assembly coupled to a proximal end of the shaft; a *battery* assembly coupled to the handle assembly; a radio frequency (RF) energy output powered by the *battery* assembly and configured to apply RF energy to a tissue; an ultrasonic energy output powered by the *battery* assembly and configured to apply ultrasonic energy to the tissue; and a controller configured to, based at least in part on a measured tissue characteristic, start application of RF energy by the RF energy output or application of ultrasonic energy by the ultrasonic energy output at a first time.

**United States Patent Application**      20210050591

**Kind Code**      **A1**

**Brewer; John C. ; et al.**      **February 18, 2021**

---

# ANODES FOR LITHIUM-BASED ENERGY STORAGE DEVICES, AND METHODS FOR MAKING THE SAME

## Abstract


A method of making a prelithiated anode for use in a lithium-ion *battery* includes providing a current collector having an electrically conductive layer and a metal oxide layer overlaying the electrically conductive layer. The metal oxide layer has an average thickness of at least 0.01 .mu.m. A continuous porous lithium storage layer is deposited onto the metal oxide layer by a CVD process. Lithium is incorporated into the continuous porous lithium storage layer to form a lithiated storage layer prior to a first electrochemical cycle when the anode is assembled into the *battery*. The anode may be incorporated into a lithium ion *battery* along with a cathode. The cathode may include sulfur or selenium and the anode may be prelithiated.

**United States Patent Application**                               **20210043941**

**Kind Code**                                               **A1**

**HORIUCHI; Hiroshi ; et al.**                         **February 11, 2021**

---

## BATTERY

### Abstract


A *battery* includes a positive electrode that includes a positive electrode current collector and a positive electrode active material layer provided on the positive electrode current collector and has a positive electrode current collector exposed portion at which the positive electrode current collector is exposed; a negative electrode that includes a negative electrode current collector and a negative electrode active material layer provided on the negative electrode current collector and has a negative electrode current collector exposed portion at which the negative electrode current collector is exposed; a separator provided between the positive electrode and the negative electrode; and an intermediate layer that is provided between the separator and at least one of the positive and negative electrodes and includes at least one of a fluororesin and a grain.

**United States Patent Application**                     20210043931

**Kind Code**                                                     **A1**

**KIM; Do-Yu**                                         **February 11, 2021**

---

POSITIVE ACTIVE MATERIAL PRECURSOR FOR RECHARGEABLE LITHIUM BATTERY, POSITIVE ACTIVE MATERIAL FOR RECHARGEABLE LITHIUM BATTERY, METHOD OF PREPARING THE POSITIVE ACTIVE MATERIAL, AND RECHARGEABLE LITHIUM BATTERY INCLUDING THE POSITIVE ACTIVE MATERIAL

## Abstract

An embodiment provides a positive active material precursor for a rechargeable lithium ***battery*** including: a nickel-based ***composite*** precursor including a secondary particle comprising a plurality of primary particles that are aggregated together, the nickel-based ***composite*** precursor having a central portion and a surface portion, and the central portion of the nickel-based ***composite*** precursor including a phosphate.

| **United States Patent Application** | **20210036368** |
|---|---|
| **Kind Code** | **A1** |
| **JIANG; Yao ; et al.** | **February 4, 2021** |

---

## LITHIUM-ION BATTERY AND APPARATUS

### Abstract

This application provides a lithium-ion *battery* and an apparatus. The lithium-ion *battery* includes an electrode assembly and an electrolyte. The electrode assembly includes a positive electrode plate, a negative electrode plate, and a separator. A positive active material of the positive electrode plate includes Li.sub.x1CO.sub.y1M.sub.1- y1O.sub.2-z1Q.sub.z1, where 0.5.ltoreq.x1.ltoreq.1.2, 0.8.ltoreq.y1.ltoreq.1.0, 0.ltoreq.z1.ltoreq.0.1, M is selected from one or more of Al, Ti, Zr, Y, and Mg, and Q is selected from one or more of F, Cl, and S. The electrolyte contains an additive A, an additive B, and an additive C. The additive A is a polynitrile six-membered nitrogenheterocyclic compound with a relatively low oxidation potential. The additive B is an anhydride compound. The additive C is a halogen substituted cyclic carbonate compound.

# I want morebooks!

Buy your books fast and straightforward online - at one of world's fastest growing online book stores! Environmentally sound due to Print-on-Demand technologies.

Buy your books online at
**www.morebooks.shop**

Kaufen Sie Ihre Bücher schnell und unkompliziert online – auf einer der am schnellsten wachsenden Buchhandelsplattformen weltweit! Dank Print-On-Demand umwelt- und ressourcenschonend produziert.

Bücher schneller online kaufen
**www.morebooks.shop**

info@omniscriptum.com
www.omniscriptum.com

MIX
Papier aus verantwortungsvollen Quellen
Paper from responsible sources
FSC® C105338
FSC
www.fsc.org

Printed by Books on Demand GmbH, Norderstedt / Germany